The Nash Styling Sketchbook

by

Patrick R. Foster

An Introductory Note from the Publisher

It is with great excitement that we introduce this new book, the first volume of what is planned to be a new trilogy. Automotive historian/author Patrick R Foster has chosen to focus his attention on the evolution of design at American Motors and it's predecessor companies. The three books that will make up this new series will collectively be called the **"History of American Motors Design."** Future volumes should include a "Rambler Styling Sketchbook" and an "AMC Styling Sketchbook." All three books in the series will feature a mix of original design renderings, drawings, sketches, photographs of concept cars, clay styling studies, dream cars, prototypes and will show some of the evolution of production car designs. It is our hope that fans of automotive history will enjoy the pictures contained in these books while at the same time gaining a deeper appreciation of the splendid work done by the designers at American Motors.

Thank you

The Olde Milford Press
PO Box 5342
Milford, CT 06460

ISBN# 0-9662019-0-6
Library of Congress Catalog Card Number: 97-095111

Cover Illustration: Nash Golden Jet rail car–design by William Reddig

Contents

Foreword

Perhaps the most surprising thing about American Motors Styling/Design history is how little we know of it. Before 1952 Nash Styling was, for the most part, seldom mentioned in the press. From 1952 and continuing for a few years the company's advertising department gave most of the credit for Nash styling to the design firm of Pinin Farina, even though Nash Styling possessed a very capable in-house design staff that did a lot of the actual work. The advertising people knew a good story when they saw it—Farina was working under contract to Nash, and Farina's reputation as probably the premier custom auto designer in the world was an asset that couldn't be ignored. Nash sometimes also employed other independent stylists for special projects.

In the mid to late 1950's some enthusiast magazines, most notably *Car Life,* did write a bit about AMC's Director of Styling, Edmund Anderson, though only briefly. It wasn't until the ascendancy of Anderson's replacement in the early 1960's, Richard Teague, that AMC Styling began to be written of in a more positive manner.

The shame of it all is that Nash and AMC designers of the 1950's never got to bask in the same sort of limelight that favored the design chief in the 1960's. Why that happened is another story, one that we may take up in the future. For now, it's better to correct past oversights by displaying herein the broad range of talent that was hard at work in the Nash design studios in the 1950's.

This book is not meant to be an all-inclusive collection of Nash designs and could not, due to the limitations of my collection, contain as many pictures as I would have liked. However, it is meant to offer enthusiasts and students of automotive history the fullest look thus far of what went on behind the closed doors of the Nash Styling studios. It has long been a dream of mine to bring together in one book a large number of design sketches, renderings, and drawings from Nash Styling, including photos of prototypes, clay models and concept cars, and the men who designed them, as a way of ensuring that the great designers and their works will never be forgotten. It's my hope that by doing so the public will better appreciate how tragic the loss of AMC really was.

There are many people that I would like to thank for their help in making this dream come true but the most important one is my good friend William Reddig. Bill has, over the course of several years, given graciously of this time helping me compile artifacts and the history of Nash Styling in the 1950's as well as its transformation into AM Design. He has also provided many of the drawings seen within. Automotive enthusiasts everywhere owe Bill a debt of gratitude for his efforts.

I also need to acknowledge the late Edmund Anderson, who donated many materials used in this book, also Bill Flajole and his daughter Diana Hawkinson, who kindly loaned materials. Thanks also to Bob Smith, Karl Harris, Bob Thomas and Bill Wilson, and to all the AMC designers whose work appears within. Special thanks are in order to the wonderful Lorenza Pininfarina for loaning us photos of Pininfarinas' Nash projects. The always generous Will Fox and Frank Peiler of Collectible Automobile magazine deserve an extra hearty thank you for loaning us many photos of Nash clay model studies. My good friend Leo Carroll deserves a round of applause for his help and advice. Thanks too, go to my brother Michael for sharing his business wisdom with me. Lastly, thanks are owed to the employees of Milford Photo for their skill in developing many of the pictures.

I've long been recognized as an automotive historian but a much more important job I have is as husband and father. I'd like to thank my darling wife Diane and wonderful daughter Caitlin for always supporting me even with my supposedly impractical projects, including two trilogies which 'experts' said were unneeded and unwanted. (I think it was Henry Ford who said "An expert is someone who will tell you what can't be done"). I'd also like to thank you, the AMC enthusiast, whose support of my books has allowed me to continue my quest for knowledge and understanding of this great industrial concern.

The photos and illustrations you will see herein represent what I believe is the largest gathering of vintage Nash design studies ever shown to the public. They are a wonderful look at Nash Styling, and many of them have never before been shown to the public. I hope you enjoy them all.

Patrick R Foster
Milford, Connecticut
March 1998

This book is dedicated to William E Reddig,
one of the great names in Nash and AMC design.
Thanks, Bill, for a job well done!

Chapter One: The Designers

In the years prior to World War Two, Nash Motors generally avoided the expense of maintaining an in-house design studio and relied instead on the services of its own Engineering Department employees plus free-lance independent stylists to design the exteriors of its automobiles. The system seems to have worked well, since for the most part Nash automobiles were sturdily built and attractively styled.

One early independent stylist was Count Alexis de Sahknovsky, credited with designing the famed special 1940 model 4081 Nash Cabriolet. With its cut down doors and windshield, this was a handsome, sporty 'play car' which unfortunately didn't sell very well. Of the twenty built reportedly only 11 found buyers. The remaining cars were converted back to their original standard condition and sold as ordinary convertibles.

Two independent designers, Ted Pietsch and Bob Koto, pitched their idea for a radically aerodynamic postwar car to Nils E Wahlberg, Nash's Vice President of Engineering sometime around 1943. Wahlberg turned down the proposal but evidently a seed was planted in his mind since the next all-new Nash design, the 1949 Ambassador and '600' Airflytes, bore a strong resemblance to what the two men had shown him. Later in this book you'll see the original Pietsch sketch and will be able to judge for yourself how much it influenced Nash styling.

Nash also had an agreement with William Flajole, an innovative independent designer who maintained a design studio in Detroit. Nash president George Mason had Flajole working on a radical new small car design for what he felt would be an emerging postwar market for 'personal use' autos, small cars designed for a specific need such as a second car for women, or as an economical commuter car.

Nash had also arranged for some proposals from a designer it had on retainer named George Walker. Reportedly, Walker had earlier designed Nash's 1939–1940 standard models (that is, all but the de Sahknovsky cars). What Walker planned for the postwar product was a cleanly styled, slab-sided sedan. Nash President George Mason held off producing it until the overheated postwar market could catch up with demand. By that time, however, Walker reportedly had peddled a very similar design to Ford, which introduced the styling theme on their 1949 models. An angry George Mason decided it was time for Nash to set up its own design studio.

Mason hired Edmund E Anderson, who had been head of GM's Oldsmobile Styling Studio before the war and who now headed up the Chevrolet Styling studios. Anderson was a capable designer and good administrator but he seems to have had an even greater genius in his ability to spot good talent. The decision to hire him on at Nash was aided, according to one source, by the fact that George Mason was friends with Anderson's parents. Mason charged Ed Anderson with establishing a Styling Studio for Nash-Kelvinator.

On the same day that Anderson started his employment another of Nash's independent designers, this one the famed Italian designer Pinin Farina, was visiting the Nash headquarters facility on Plymouth Road in Detroit. The two great designers met and shook hands in George Romney's office. They were to work together on new Nash designs, the first of which was the 1952 Nash senior line. Although Nash advertising emphasized Farina's name and reputation, in truth the actual production cars were a composite of Farina's and Anderson's ideas. Farina acknowledged that himself when he referred to the cars as " . . . our jointly designed 'Golden Airflytes'" in an open letter to his Nash associates. Nash stylists claimed that they performed the bulk of the design work. The 1952 Nash production cars do, however, clearly show Farina's influence, particularly in the reverse slanted "C" pillars.

One of the first designers Anderson hired for the new Nash Styling team was Bob Thomas, a man with great talent as an artist, and who later enjoyed a very successful career with Ford. Also hired on were stylists Donald Butler and Royland Taylor, as well as Charl E Green, an artisan particularly skilled in clay modeling.

Of the men who joined up with Anderson, Bob Thomas, Don Butler and Royland Taylor all contributed greatly to the look of Nash (and later AMC) designs. But to Ed Anderson, one man in particular stood out. In later years Anderson would remember that man, as "the best—the very best." That esteemed designer was the youthful, enthusiastic Bill Reddig.

Reddig had started out with Ford, signed on with Anderson as a stylist at Nash and arrived in time to work on a great project, the all-new 1952 Nash Statesman and Ambassador lines. These were the cars the company called the "Golden Anniversary" Nashes, in recognition of its 50 years in business. Reddig later worked on the top selling Nash Rambler four door sedans and station wagons that debuted in 1954, as well as the Metropolitan small car that also debuted that year. Reddig was heavily involved in the design of the landmark 1956 Rambler and the original Ambassador by Rambler that arrived in 1958. After the formation of American Motors, Jack Garnier was made Nash studio chief and Reddig was elevated to Assistant Director of Styling.

Many of these great designers appear on the next few pages.

The legendary George W Mason, Chairman and President of Nash-Kelvinator, was instrumental in having the company set up an in-house Styling department.

Edmund Anderson, formerly a General Motors design executive, was hired by Mason to head up the new Nash Styling department.

Anderson hired William E (Bill) Reddig away from Ford. Reddig soon became Anderson's right hand man. Anderson remembered Reddig as " . . . the very best."

Recognized throughout the world as the leading custom car designer of our time, Pinin Farina and his cars have been honored with more than a hundred Grand Prix *in international competitions.*

Pinin Farina was also signed up by Nash to help style cars. Farina had a strong influence on the 1952–1957 senior Nash line, the 1953–1955 Ramblers and was solely responsible for the gorgeous 1952–1954 Nash-Healey series.

Charl E Greene, master clay modeler for Nash, puts the finishing touches on a scale model of the Nash Metropolitan.

Another independent designer for Nash was William Flajole. Flajole designed the original prototype of the car that became the Metropolitan.

The Nash Styling team in the early 1950's. The photo is dated August 23, 1951. The men, left to right are: Styling chief Ed Anderson, designers Bob Thomas, Royland Taylor, Bill Reddig and Don Butler. They are standing in front of a full sized rendering of the proposed 1952 Nash senior line.

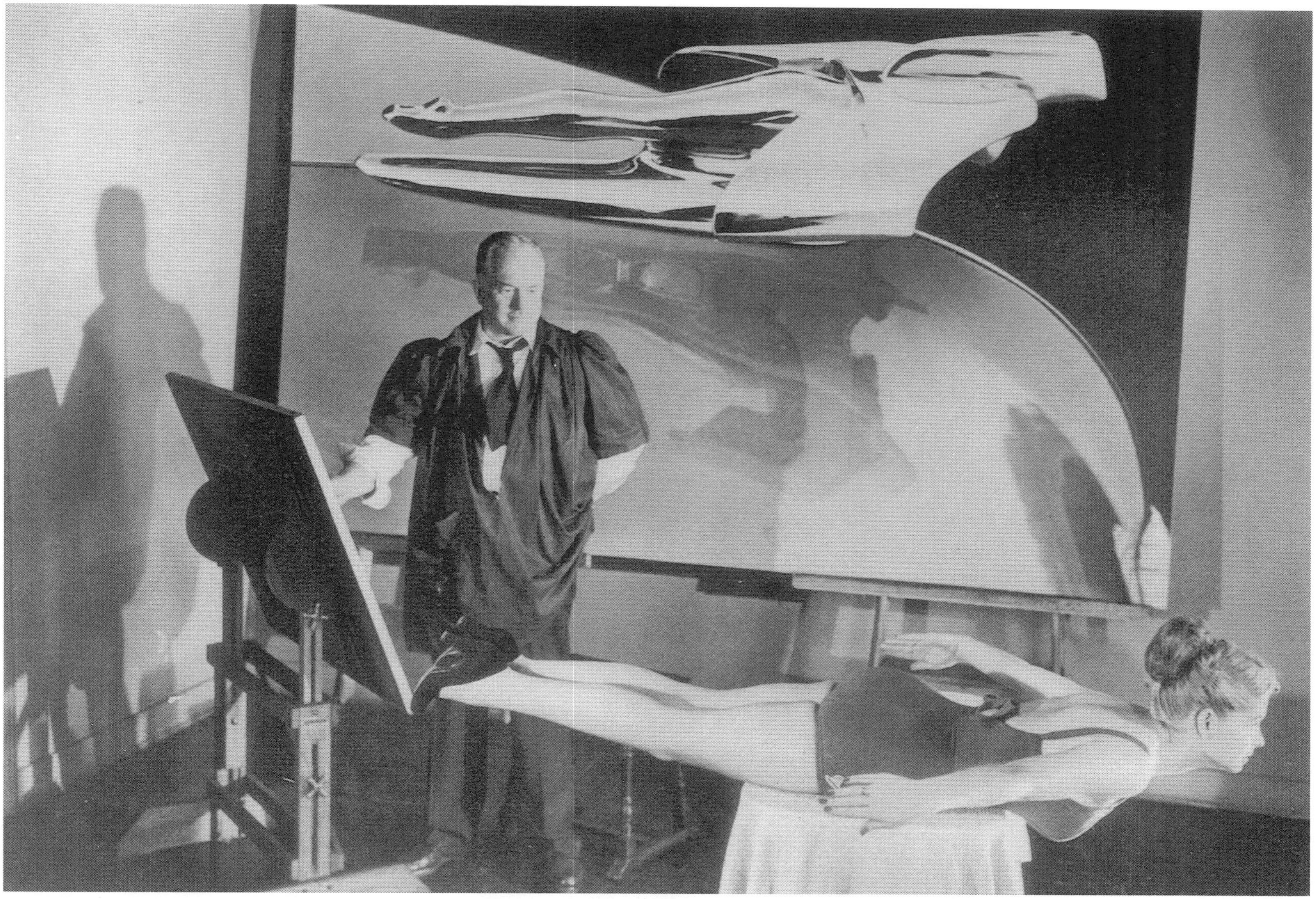

George Petty, who designed those famous Nash hood ornaments, evidently liked to work from full sized models.

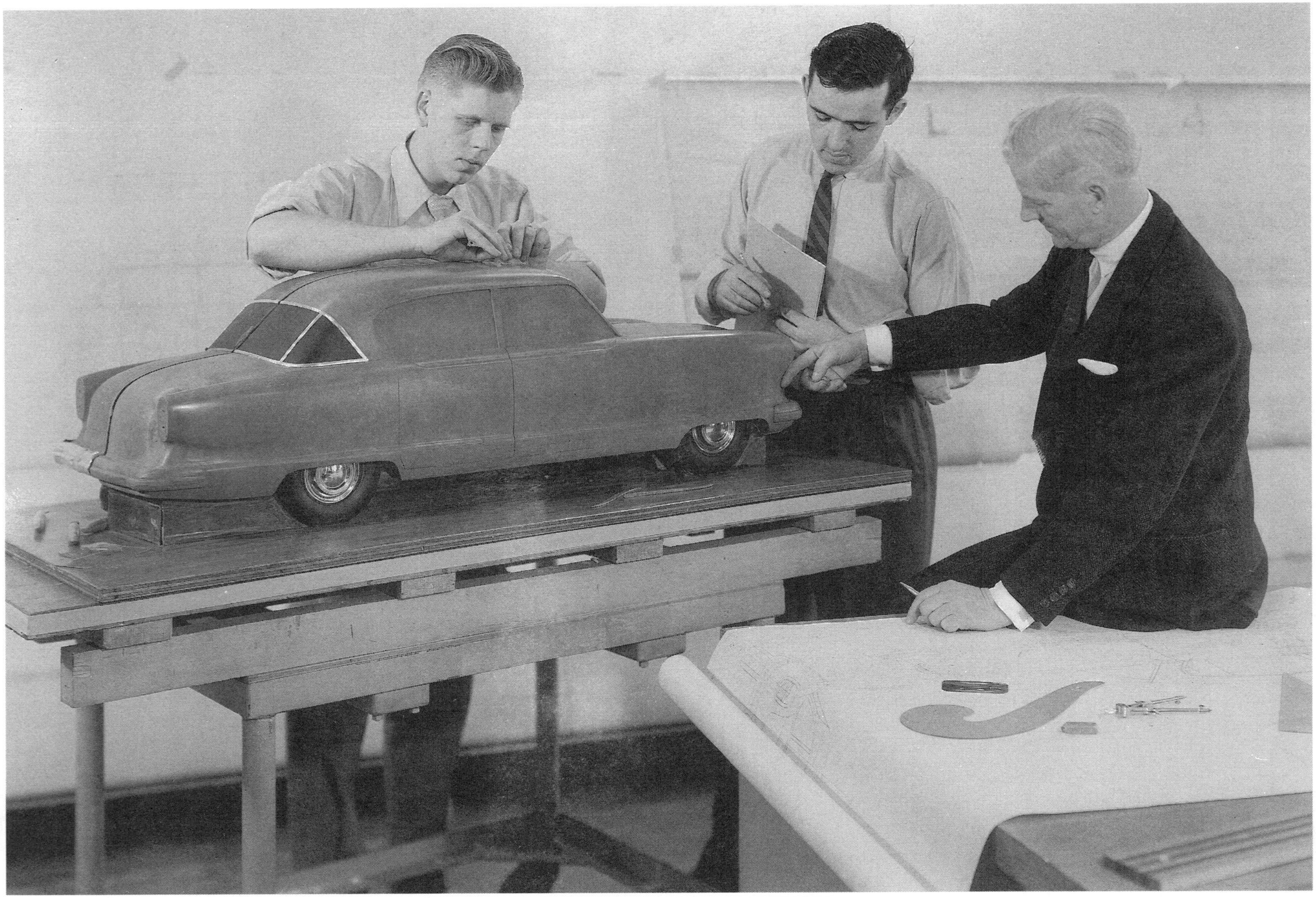

An unidentified modeler works with Reddig (center) and Anderson (right) on detail changes to the 1952 Nash senior line.

Ed Anderson, in dark suit, points out details on a mock-up of a new grill design that appeared on the 1954 Nash senior line. Two men bending over drafting tables in the background are Bill Reddig (left) and Charl Greene (right).

Anderson inspecting a wheel cover design with accessories designer Don Butler. Name plates mounted on the wall read 'Nash Embassy,' 'Nash Rambler Lake Shore' and 'Rambler Debutante.'

Anderson looks over design proposal with Jack Garnier. After the merger of Nash and Hudson, Anderson set up separate Styling studios for each. Garnier was in charge of the Nash Styling studio, evidently being the last man to hold that spot.

American Motors styling executives circa 1955. Left to right they are: Don Stumpf, Interior Studio manager, Don Butler (standing) Accessories designer, Ken Samples, Hudson Styling Studio manager, Anderson, Bill Reddig, Assistant Director of Styling, Jack Garnier, Nash Styling Studio manager.

Stylist Count Alexis de Sahknovsky, shown at the wheel of a special Nash model he created in 1940.

Chapter Two: Designs for the Postwar Nash Senior Line

The story behind the Nash Golden Anniversary senior line has already been told many times. Anderson and his team of Nash stylists had worked diligently on a handsome and elegant sedan to debut for 1952—which was Nash's Fiftieth Anniversary. Just as Anderson was signing off on the design, word came down to halt all work on the project. Mason, influenced by Nash Engineering chief Meade Moore, wanted Pinin Farina in on the job. Anderson's team was disappointed but the men accepted the management dictate. Farina built a full sized steel prototype that was shipped to America, where Nash executives eagerly awaited it. When it was unveiled, however, the men came away somewhat disturbed. Although it showed many good features, it was felt that overall Farina's design was too "European" for American tastes. However, Mason decided to utilize the best features from both designs, so Anderson was instructed to have Nash Styling consolidate the two design themes, his and Farina's, onto one model, which he later referred to as "a composite". The resulting design won acclaim as one of the best looking cars on the market. The blending of the European and American ideas resulted in a car that looked at home anywhere in the world.

What follows are several pages that show various designs for Nash's postwar senior cars, beginning with several clay models that illustrate several different design themes that Nash explored. The chapter includes drawings of what became the 1952 senior line, as well as some interesting rough sketches of different directions Nash 'big car' styling might have taken. We've included some unique photos of production cars as well. Shown too are pictures that depict where senior car styling might have evolved if Nash had remained in production.

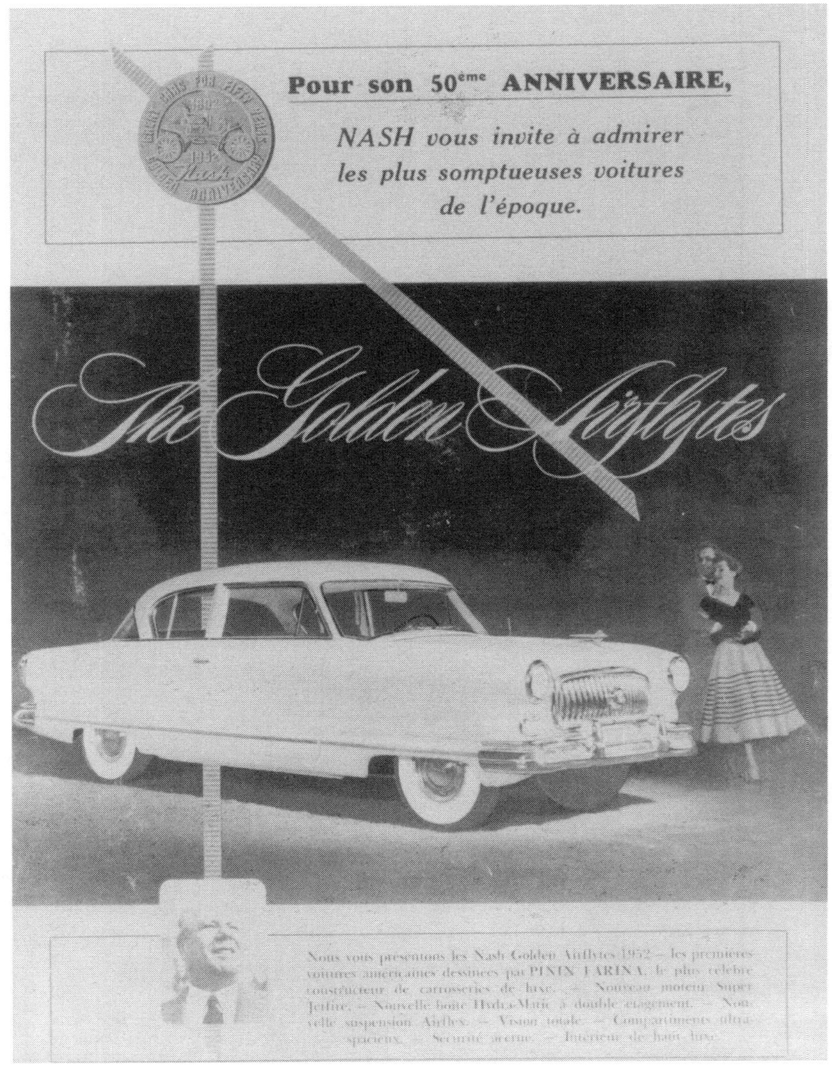

The sales brochures for the 1952 models, including this export version, extolled the beauty of Farina design.

This sketch by Ted Pietsch was shown to Nils E Wahlberg of Nash Engineering sometime around 1943. Wahlberg evidently liked the idea, since production Nash cars of 1949–1951 looked very similar to this concept.

Compare Pietsch's drawing with this ad for the 1949 Ambassador.

Early 1950's drawing by Pinin Farina shows handsome, low slung Ambassador. Note the non-wraparound windshield.

Two views of a Nash design study. This small clay model shows a design direction Nash didn't take, though grill hints of NXI experimental car.

Side view of small clay model shows Nash trademark enclosed wheels.

This drawing, artist unknown, shows one stage in the evolution of the 1952 styling. In this picture the grille is beginning to take shape, and the front fenders are similar to production models. The roof line still has a long way to go–the reverse sweep 'C' pillars haven't shown up yet.

Two views of clay model shows how body design is still evolving. Pointy tail lights look odd, as does the extreme body sculpting.

In these photos the grille is just about set, but fenders and roof line are still in progress.

Close now to the actual production cars, these photos clearly show how the senior Nash would have looked without reverse sweep 'C' pillars.

More views of various ideas for big Nash. Lower photo appears to have strong European influence.

Side view of proposed 1952 Nash by Nash Styling studio.

This photo shows the original final design by Nash Styling for the 1952 senior cars. This is the car Anderson planned for production before orders came down to incorporate Pinin Farina's ideas into a consolidated proposal which became the actual 1952 production car. Nash Styling employees left to right are Bob Thomas, stylist, Don Butler, stylist, Bill Reddig, stylist, Bill Fox, modeler, next man is unidentified, Ray Smith, stylist, and Ed Madison, hardware craftsman.

Front view of Nash Styling proposal shows elegant design touches, grill somewhat similar to production car.

Rear view of Nash Styling proposal shows handsome, understated approach.

Another side view of Nash Styling mock-up.

This is believed to be a production car, possibly photographed in Italy.

Nash took advantage of Farina's fame as a custom car designer by using Farina's name and picture in Nash sales catalogs, print advertising, and outdoor billboards.

Sketch by Bill Reddig shows one idea for fifties senior Nash line, and features Nash's style-mark fully enclosed wheels. Note that the roof line is similar to the Metropolitan hardtop.

Clay model shows early version of Nash design for 1955. The grill shape is set, the beautiful rounded fenders are just about right too, but notice the wraparound windshield hasn't shown up at this stage.

Front end design is clean, dramatic. Note the different shapes of the running lamps, as well as the chrome trim on the left fender.

Side view of proposal for the 1955 Nash.

Pinin Farina offered this prototype as his idea for the 1955–56 full size Nash. It seems to have influenced design of Nash production cars for those years. This car still exists.

Front view of Nash clay model clearly shows the influence of Farina prototype.

Earlier version of Farina design proposal wears plain hubcaps and appears to have a split windshield. Overall theme shows an evolution of 1952 styling, including open wheels.

Pinin Farina built full sized running model to show off his ideas for future Nash cars. Original photo caption from Farina files is marked "Nash Ambassador Berlina 1955".

Sketch by Bill Reddig has two different front fender treatments, substantial tail fin.

Similar sketch by Reddig now shows quad head lamp treatment.

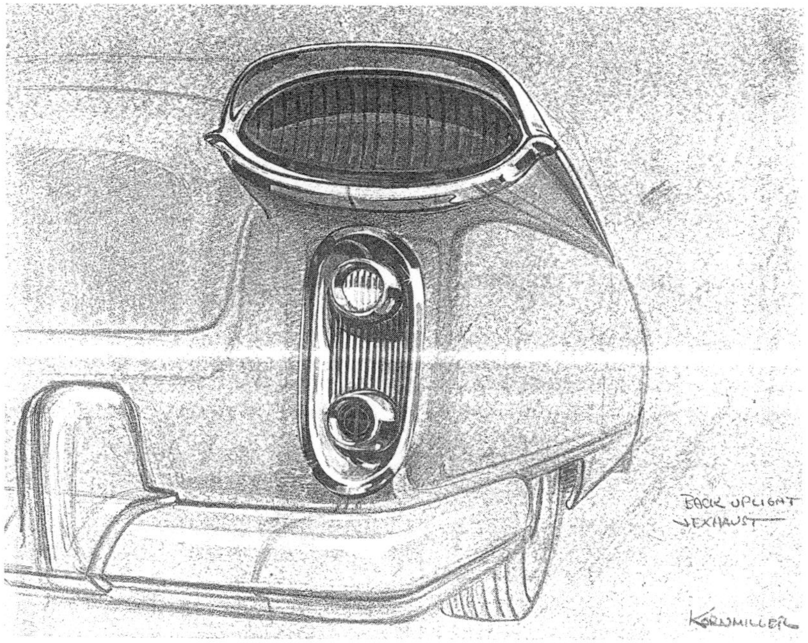

Detail of tail lamps was drawn by stylist Alan Kornmiller.

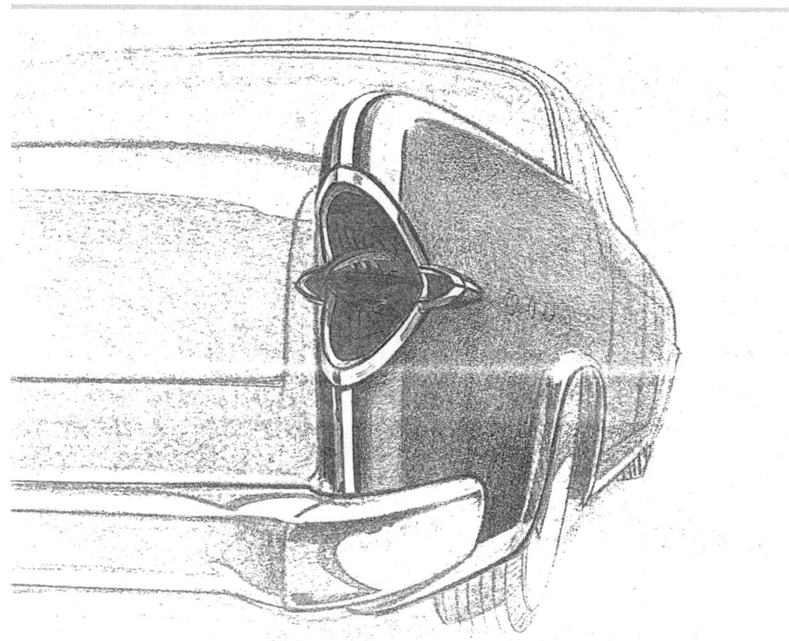

This sketch is unsigned but may also have been done by Kornmiller.

In the fifties, stylists were challenged to figure out the best way to integrate quad head lamps that would first appear in 1957. In this and several other sketches Reddig tries different approach to the problem.

Side view sketches show more up to date look for Nash-open front wheels, semi skirted rear wheels, moderate up kick to rear fenders. The overall look is very similar to late fifties Chrysler products.

Rough sketch by Reddig shows dramatic fin, very interesting head lamp and grill shape.

Sporty car has semi enclosed front wheels, canted fins.

Chapter Three: Nash-Healeys That Might Have Been

It doesn't appear that Nash ever seriously considered designing or building the Nash-Healey sports car in-house. However, Ed Anderson loved the Nash-Healey and his designers knew it. Anderson himself owned at least three N-H's, including one that he stuffed a V-8 into, and one fitted with a Hudson Hornet six cylinder racing engine. Anderson's main job was to oversee the styling of Nash (and later AMC's) line of family cars. Pinin Farina was the man who was assigned sole responsibility for the sports car designs. But Anderson and his men sketched and perhaps sometimes dreamed that they might get the next assignment to style a Nash sports car. That didn't happen but a few of the designers ideas have survived and can be seen on the pages that follow.

In 1997 we asked Bill Reddig if the designers in Nash Styling had ever been asked by management to submit designs for a Nash-Healey. Here's what Bill replied:

> *"That was not openly discussed. . .Mason was really reaching out to get the best of both worlds whether it was English or Italian or whatever. Ed Anderson wanted [us] to have a shot at [styling] it, so I wasn't about to be shy.*
>
> *Anderson wound up with three Healeys. . .shoehorned a leftover Hudson Hornet stock car racing engine in one, and a V8 in another. He drove 'em like crazy. But you see, in my fooling around with those [Nash-Healey concepts] there were certain ideas, things you come up with that perhaps could be used on [the regular Nashes]."*

The following pictures are probably the most comprehensive look yet at Nash-Healeys that might have been.

This photo, provided by Pinin Farina, shows a Nash-Healey with what appear to be unique styling embellishments—two small blisters that flank the hood scoop, and another smaller scoop mounted atop the regular scoop. This may be a production prototype or simply a production car that was built slightly different from others.

The all-new 1952 Nash-Healey as displayed at an auto show. That's Pinin Farina standing next to a picture of himself. The other gentleman appears to be Nash executive vice president George Romney.

This photo is one of the rare Nash-Healey 'LeMans' coupes. Elegant design is very European, still looks beautiful today.

Shown here is a Nash-Healey hardtop with the reverse sweep rear pillars as used on the final series of Nash-Healeys.

Bill Reddig sketched this racing version of a Nash-Healey proposal sometime in the mid 1950's. Note rear roll bar is similar to the basket-handle 'C' pillars used on the 1956/57 Rambler, though forward slanted rather than reverse.

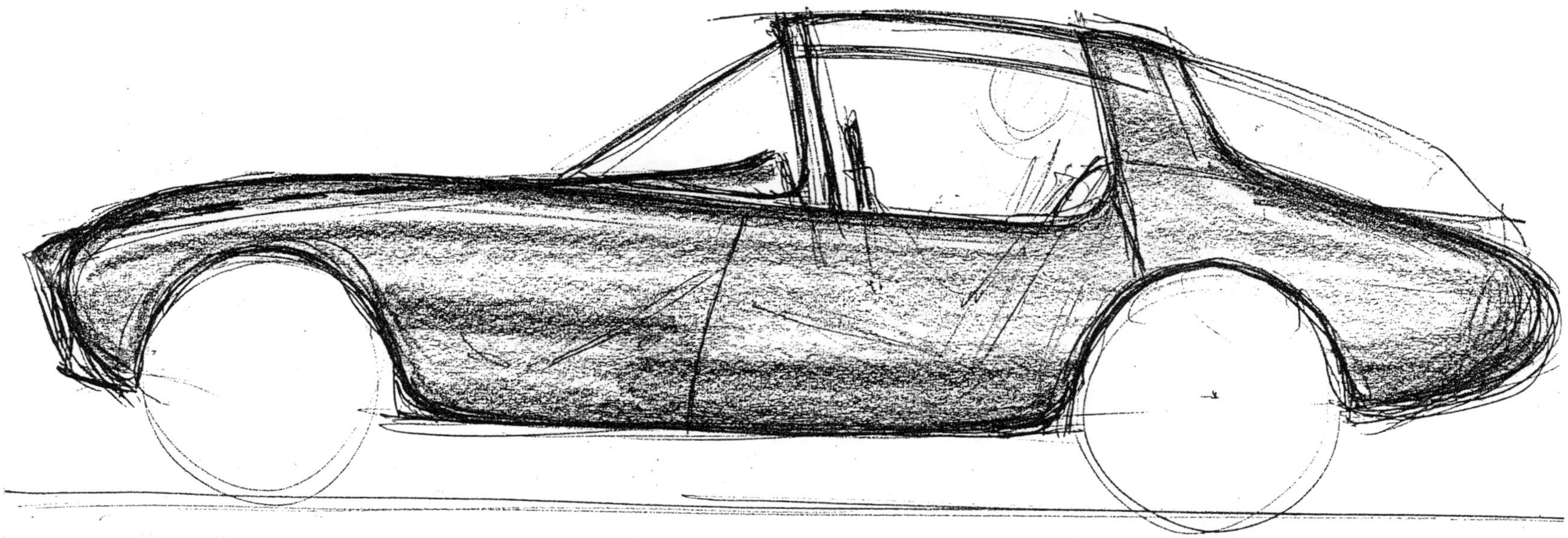

Very rough sketch by Reddig shows styling theme carried one step further to include glass roof and fastback rear window. Despite being a fifties sketch this car looks surprisingly modern.

Dramatic painting by Reddig, this shows front end of a Nash sports car concept. Very interesting head lamps and fender shapes.

Beautiful painting by Reddig of a Nash Competition Sports car. This picture is dated 8/14/51.

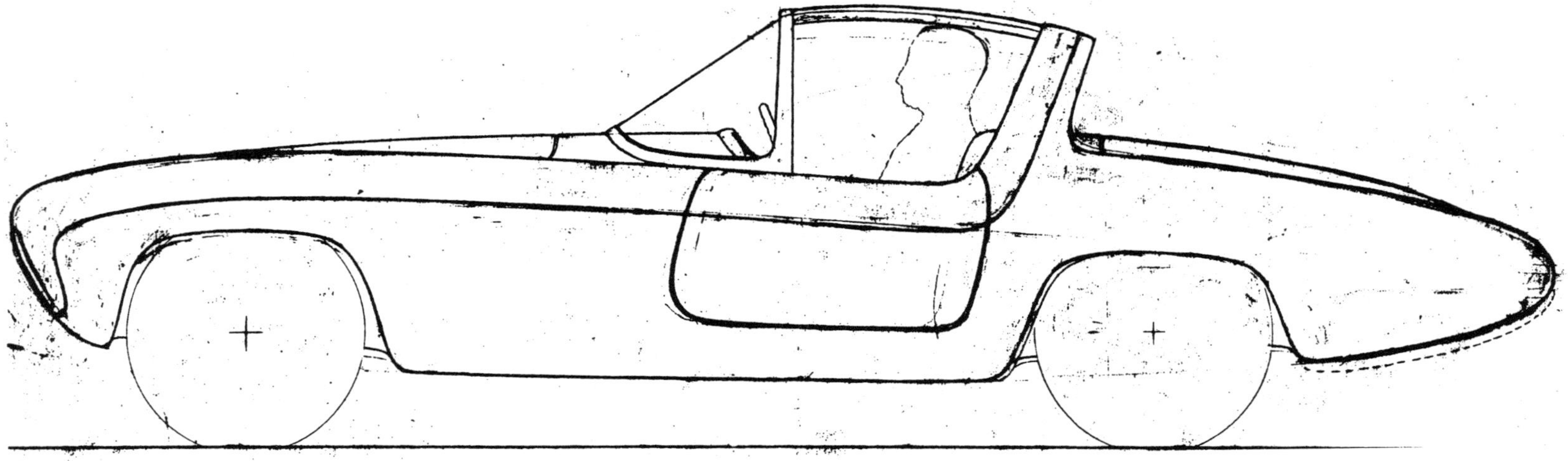

A new look for the Nash-Healey is seen in this sketch by Bill Reddig. Here the roll bar is reverse slanted like Nash passenger cars.

Simple sketch by Reddig manages to capture the romance of sports cars.

Freelance designer Bill Flajole designed and built this custom Nash-Healey roadster, based on a 1951 model. This car was given to baseball great Ted Williams as part of a promotion by the makers of Johnson Wax.

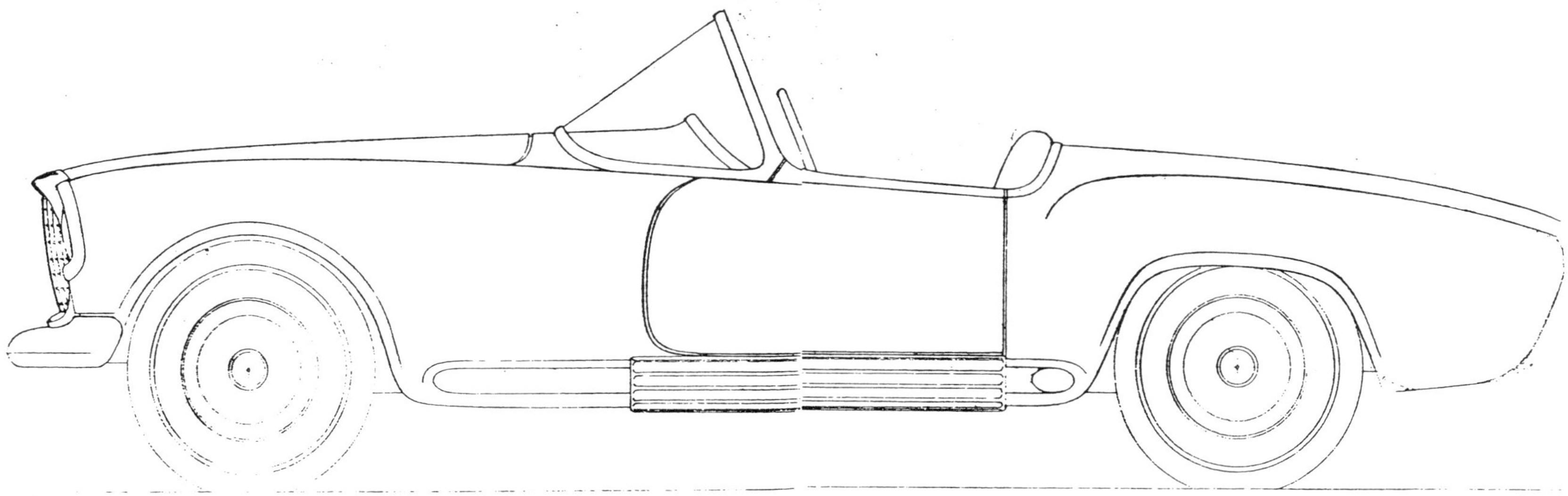

Another sketch by Reddig, this shows a Nash-Healey with a distinctly European look, has side pipes but no roll bar.

If Pinin Farina had built Nash-Healeys in the later part of the 1950's, they might have looked like this one-off prototype dubbed the Palm Beach.

When Farina first pitched this idea to AMC the car wore Nash wheel covers and a Nash hood ornament, as shown in this photo. Later, the car was referred to as the *Rambler* Palm Beach and was fitted with Rambler wheel covers.

The artist is unknown but the car wears the nameplate 'Wisp.' Unfortunately, the car is strictly a drawing, was never built. Front fenders have coves much like later Corvette would wear. The photo of the attractive lady was supposedly clipped from a magazine and added to the sketch to lend some realism to the picture.

Another painting by Reddig, this shows a more modern design. Note the tops of the wheels are slightly hidden by fenders, a variation of the traditional Nash look.

This rather large looking sport coupe was sketched by Nash's Jack Garnier. While the grille and roof line recall the 1956 Rambler, this car would have been much sportier.

Chapter Four: Rambler s By Nash

When we think of Nash we often also think of the Rambler and in some peoples minds the two names are linked so closely that seemingly one cannot be said without the other. And yet for most of Nash's history there was no Rambler in the line-up. The predecessor of Nash Motors, the Thomas B Jeffrey company, had marketed its earliest cars as Ramblers, from 1902 to 1913, but then switched the brand name to Jeffrey. That was the brand name until Charles Nash decided to use his own name on the company he bought, renaming the car line beginning in 1917. Although the Rambler had a worldwide reputation as a quality car Nash chose not to use that name again until his successor, George Mason, called his new 100 inch wheel base compacts Nash Ramblers when they debuted in 1950.

The new Rambler shared nothing with the old models except for a legacy of quality construction and superb durability. The first of the modern Ramblers were a bit tall and had a high blunt front end but the overall design was handsome. Nash Styling, and later AM Design, with input from Pinin Farina as well, worked progressive face lifts on the car beginning in 1953, over the years lowering the hood, fenders and roof, and modernizing the design. They did good work, too—the Rambler grew more handsome year by year. The basic chassis continued clear through to 1963 as the Rambler American.

Thus far only a handful of Nash Rambler designs have landed in my collection and the few I have are shown herein.

Front end detail by Jack Garnier shows development of what became the 1956 Nash Rambler.

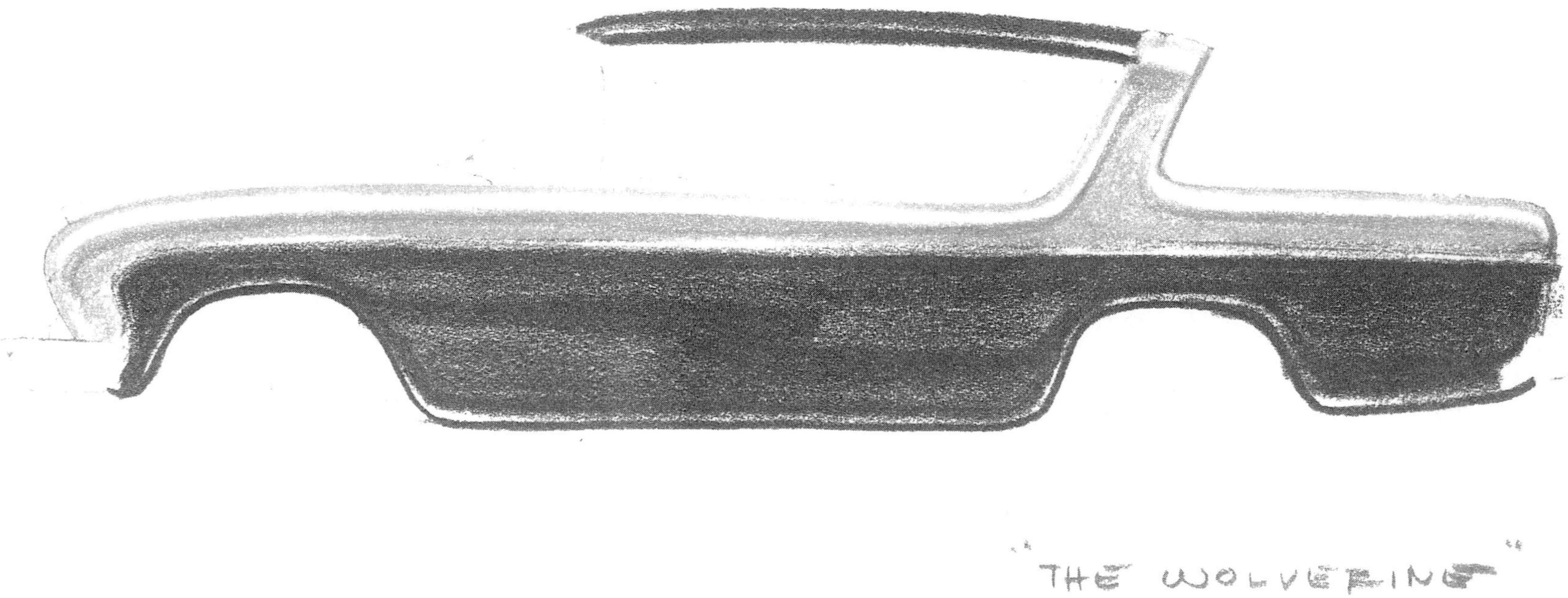

Side view sketch by Bill Reddig, titled "The Wolverine," suggested a possible two tone paint scheme for the 1956 Nash Rambler.

This lovely photo shows a Pinin Farina prototype of the 1953/54 Rambler. Note the unusual front wheel cut out, somewhat like later 1955 Nash senior cars. Also note front bumper guards, unique chrome trim on body sides and roof.

Sketch by Reddig depicts Rambler four door hardtop with open wheel wells, louvered hood.

Another Reddig sketch shows variation on the Rambler hardtop theme, with thinner bumper, lower hood line, small tail fins.

Chapter Five: Those Wonderful Metropolitans

When Nash-Kelvinator introduced the sprightly little Metropolitan as a Nash product in March of 1954, it was entering nearly uncharted waters. The tiny Crosley car, first offered prior to World War II, had tried unsuccessfully to compete in the postwar market. At first, Crosley sales were fairly good, helped no doubt by a national shortage of cars of any type. But when the market evolved from a sellers to a buyers market, sales of the diminutive Crosley dried up.

The Metropolitan thus entered a market that was very competitive and it would be forced to win each and every sale on its merits. Luckily, the car had been well designed by Nash Styling.

Most of the credit for the Metropolitan belongs to William Flajole, an independent stylist with a studio in Detroit. Flajole had come up with the basic concept, pitched it to Nash CEO George Mason and won a design contract from Nash. Flajole went on to refine his design and build a full size prototype, which toured the auto show circuit to test public reaction.

The results were encouraging and Nash decided to commit to full scale production. The job of refining the car for series production fell to the regular members of Nash Styling, Bill Reddig, Bob Thomas, Royland Taylor, Don Butler and Charl E Greene, headed up by Director of Styling Edmund E Anderson.

The pages that follow show various design proposals by Flajole and the Nash Styling designers involved with the project.

Early sketch by Bill Flajole shows basic concept of car: two doors, two passenger seating, plenty of style. Note head lamps set into frontal area, not into fenders. This car bears the nameplate "Fiat Esquire" to camouflage its real corporate parent.

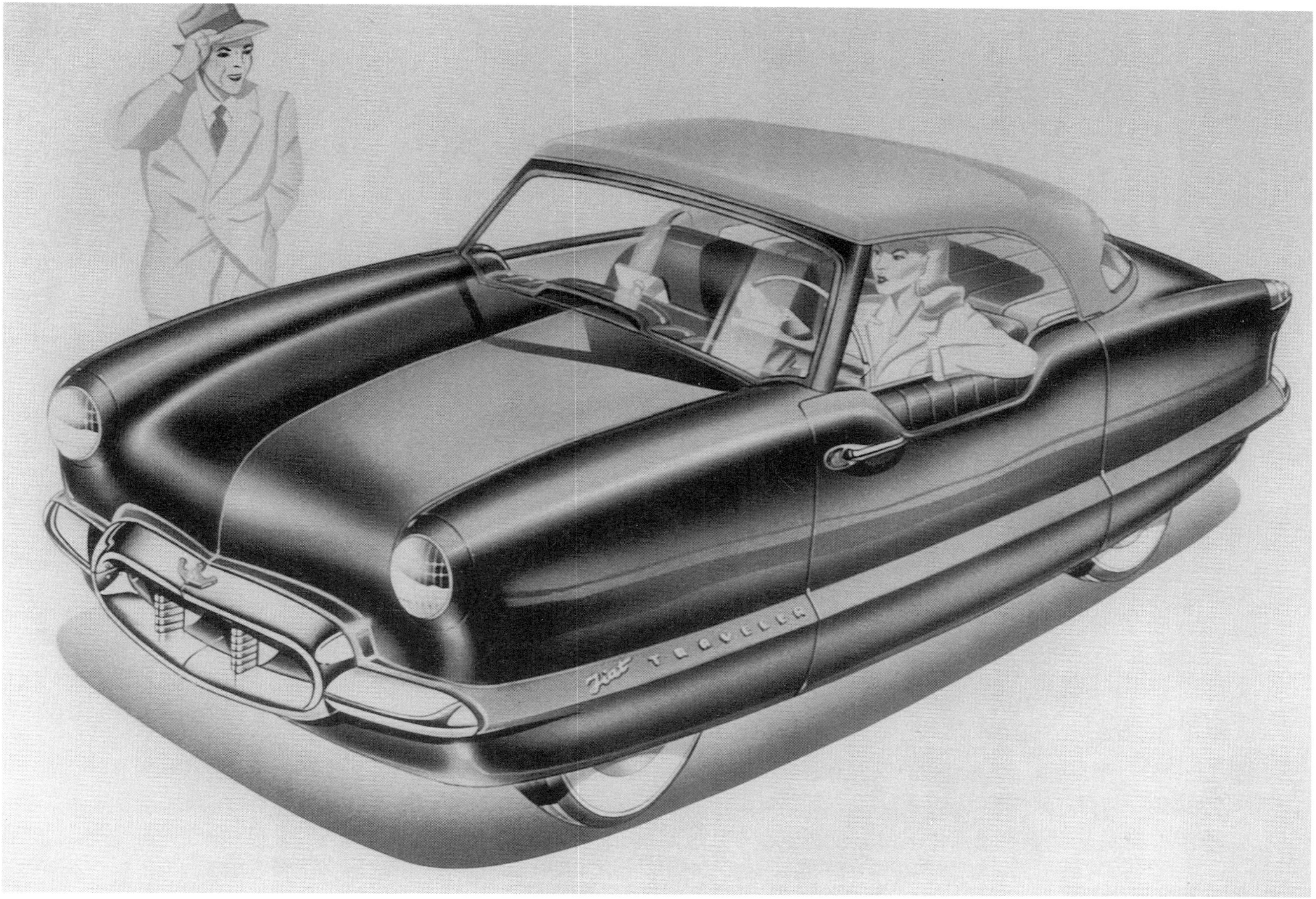

Another drawing, this car has the beginnings of Flajole's one piece bumper/grille idea and now has conventional head lamps set into rounded fenders. Note suicide-style door hinging. Windshield drops down at the corners to meet the cowl. This car bears the name "Fiat Traveler," again as disguise.

This concept drawing shows a small car with station wagon style roof line, interesting split rear window. Note how the lower body lines blend into the rear bumper.

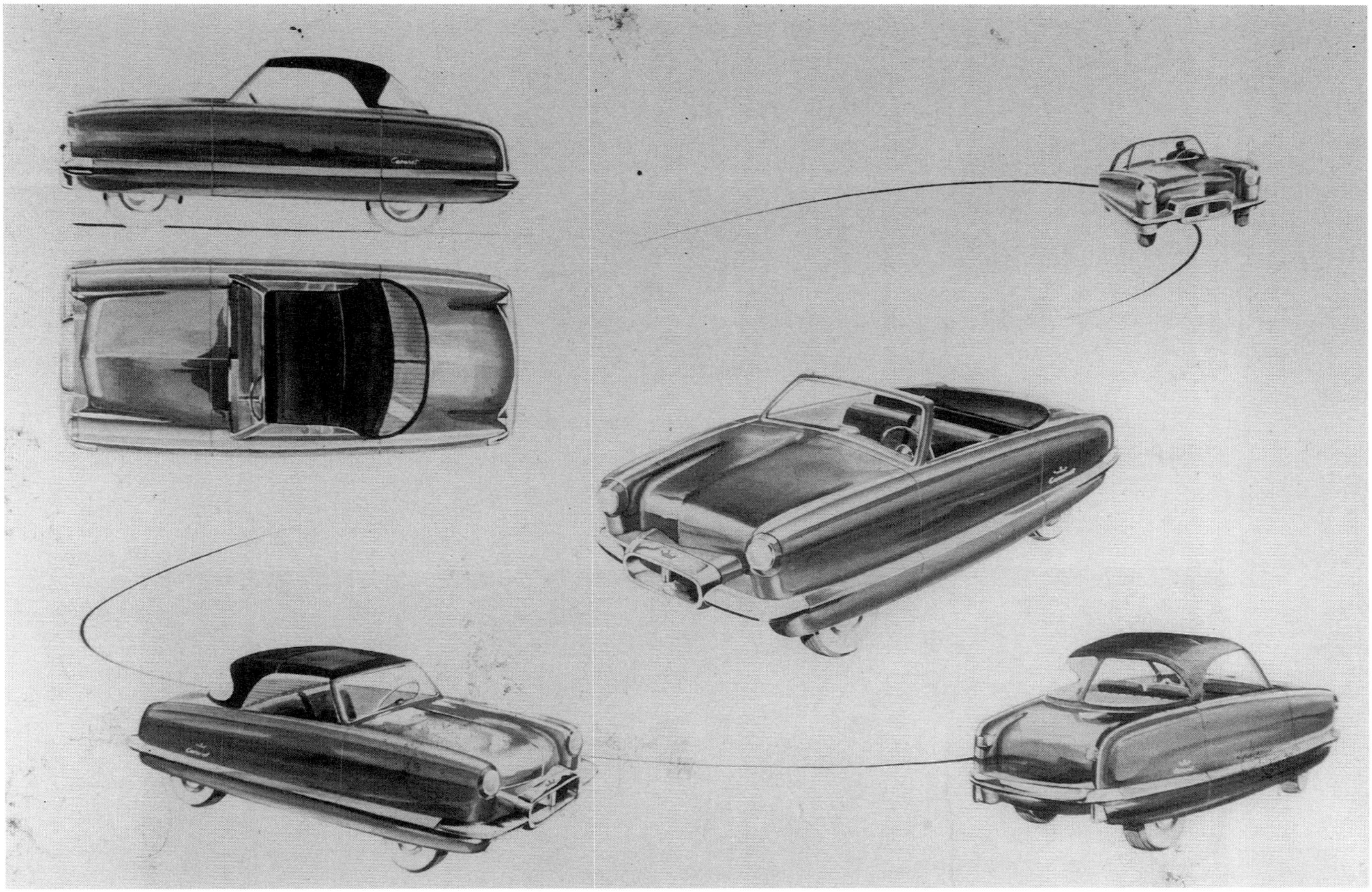

Different views of Flajole's small car concept.

Flajole sculpted this clay scale model of his concept car, was given the go-ahead by Mason to build a full-size running prototype.

First running prototype of what became the Metropolitan. This photo was taken right after car was built but probably before the interior was installed. Designer William Flajole is shown at the wheel.

The Flajole prototype, dubbed the NXI (for Nash Experimental International) soon had an interior installed by a separate company. After the body was painted a deep maroon, and wide white wall tires and full wheel discs were installed, the car was displayed at several auto shows. In this photo we see George W Mason, chairman of Nash-Kelvinator, at the wheel, with his executive vice president George W Romney seated next to him. Both men were confirmed small car enthusiasts.

After displaying the NXI, a series of more advanced prototypes was built. These cars, called NKI (for Nash-Kelvinator International) were different from the earlier NXI. The hood now wore a prominent blister with a chrome scoop at the forward end, the rear fenders were semi-skirted instead of fully skirted, and fresh air ventilators were set into the front fender sides. The car also has turn signal lamps, which the NXI lacked. In this photo Meade Moore, VP of Nash Engineering, stands next to an NKI hardtop prototype. Inside the car is George Mason at the wheel. The other passenger is unidentified. Other body styles in the NKI series were a two door convertible and a two door coupe, but no photos of these have surfaced thus far. A sharp eye can spot another NKI prototype lurking in the shadows behind Moore—the grille is evident in the far left of the photo just ahead of the first car's head lamp.

Photo of a full size drawing of the Metropolitan during development of styling for the production version. Compare this with next photo to see how the design looked when rendered in metal.

This photo shows a pre-production prototype Metropolitan. Non-production features on this car include one-piece rear window (which showed up on 1958 models), full wheel discs, 'Custom' nameplate.

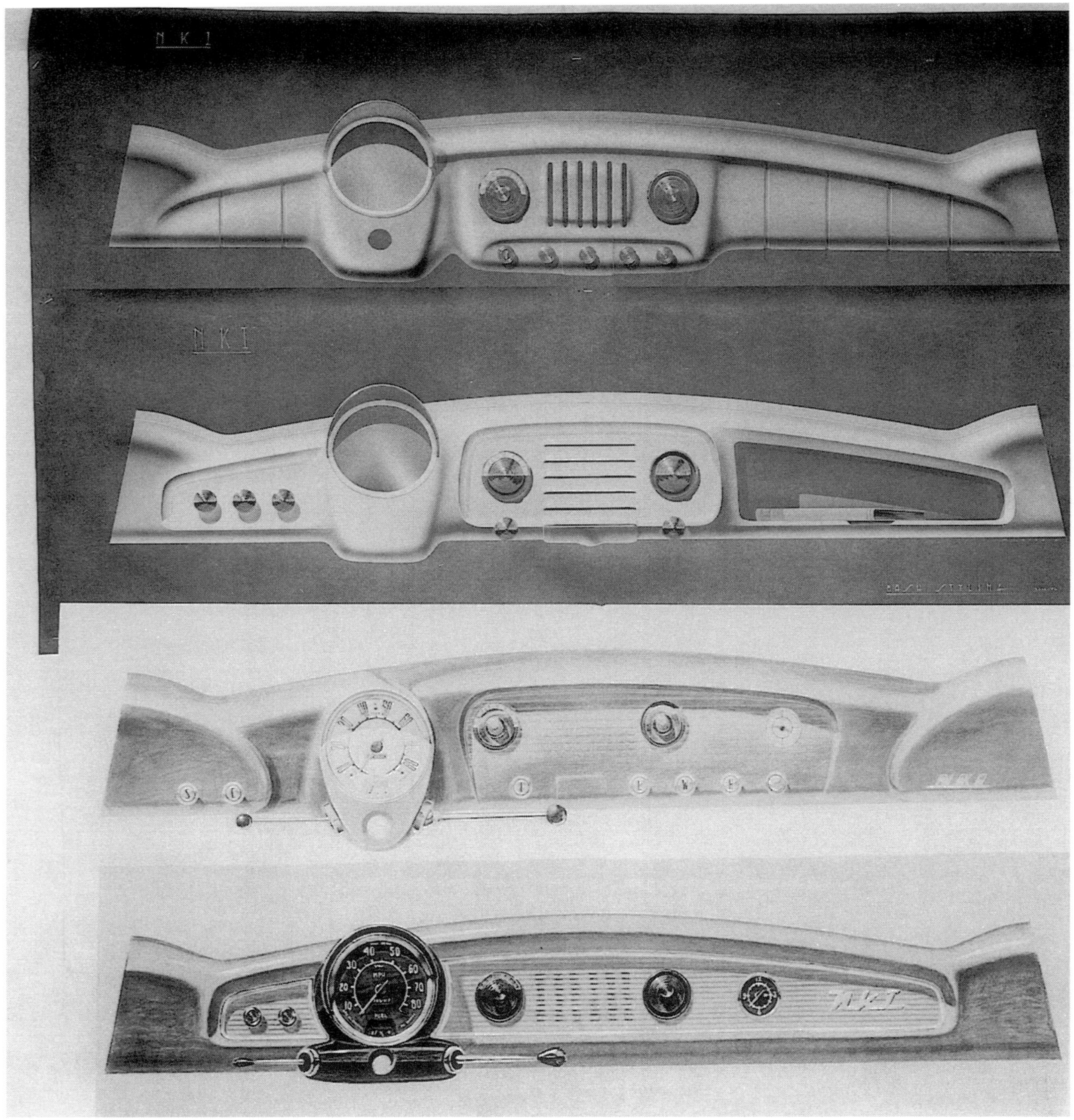

Four different Metropolitan instrument panel designs by Bill Reddig. The one second from the top was chosen for production models.

Rear view of Metropolitan station wagon prototype designed by Nash Styling. Note the sliding rear side windows. Car also appears to have a flip open rear window but rear mounted spare indicates lack of an opening tailgate.

Front view of Met wagon shows interesting hood ornament, revised side moldings and sliding side window in the open position. The roof appears to be a darker color than the body.

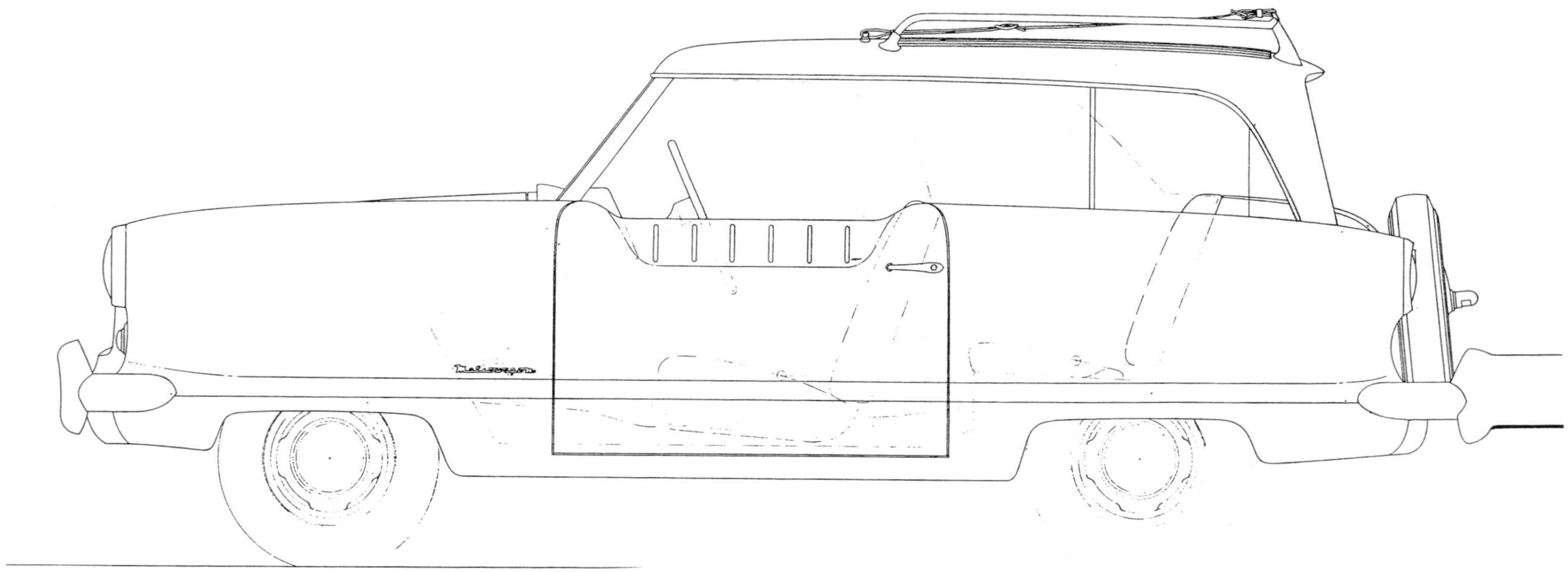

Here is an original drawing by Bill Reddig of the proposed Met wagon. Note how Reddig cleverly included a ghost outline of a Met coupe roof line to illustrate how much more interior room could be had in wagon. Seat positions also appear in outline form.

Pinin Farina also built a Met wagon. Note this prototype wears the Nash grille badge, indicating the car was based on a modified 1956 Met.

Side view of Farina's Met proposal shows handsome style, unique 'B' pillar and roof line.

Rare look at the back of Farina Met wagon shows the cabin had light, airy feel, with excellent visibility in all directions. Note the opening tailgate, no outside spare tire. The rear window appears to be fixed, but rear side windows appear to be sliders.

Chapter Six: A Look at the Future!

Perhaps the single greatest aspect of life in the 1950's was the public's unbounded enthusiasm for the future. In magazines, newspapers, radio, television and in the movies, the message was clear: all that was needed to overcome any inconvenience of life was a solid measure of space age technology. There was talk that nuclear power plants would produce household electricity for so little cost it would be "too cheap to meter." In the future, everyone would live in a home like the Jetson's, take vacations on the moon and live in unabashed, futuristic affluence, all provided by miraculous new wonder machines.

Automotive designers caught the publics enthusiasm and very soon rocket and jet powered 'Cars of the Future' began to appear in popular magazines and on the stands at the larger auto shows. Nash, of course, as America's most innovative car company, always had an eye to the future—and in the styling studios on Plymouth Road in Detroit the designers sketched out their vision of what the future would hold for personal transportation. To these men the future looked dazzling! The front cover of this book features Bill Reddig's wonderful Nash Golden Jet rail car, a design we feel is the very epitome of the exuberant decade of the 1950's in America. Pictured here are several other surviving examples of what Nash designers had in mind for the coming years.

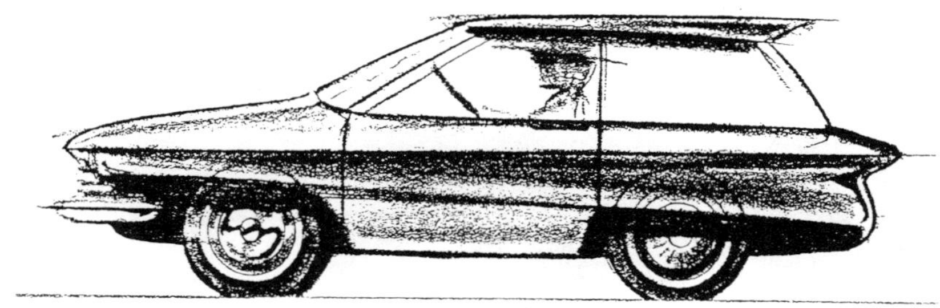

In the mid–1950's Bill Reddig envisioned this futuristic Metropolitan. Despite the passage of 40 years, this design still looks fresh and modern.

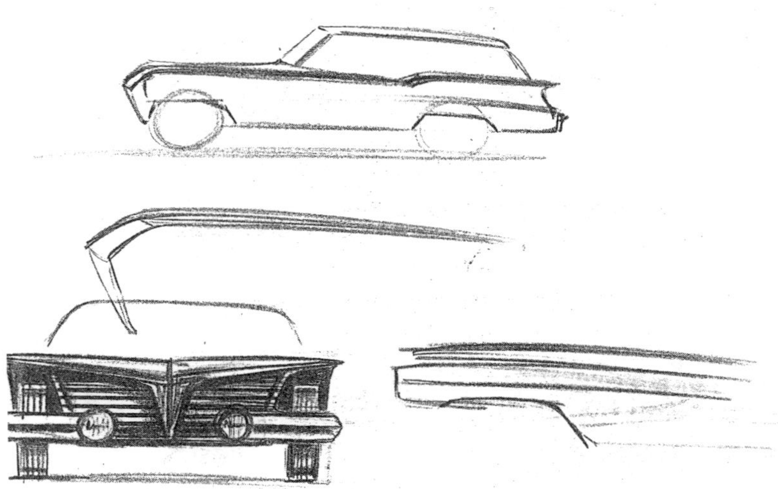

Reddig sketch shows how designers try different ideas out in rough drawings.

Design sketch for a senior Nash has aggressive modern looking style.

Another look at Reddig's Golden Jet rail car, this version wears the Nash grille badge and Nash name on front fender.

Another Reddig sketch of updated Metropolitan shows split bumper, hooded head lamps, and model name shortened to 'Metro.'

Rare sketch by Reddig shows Met with finned head lamp surrounds, centrally mounted radio aerial, and T-top roof—many years before the T-Top was first offered on a production car. Note how the interesting bumper blends in with the side body sculpting. Note, too, the Nash emblem inset in grille.

Reddig sketch of big car shows quad head lamps set into grille area, Dagmar-style bumpers.

Futuristic van concept by Reddig dates, amazingly, back to the 1950's.

Several doodles by Reddig illustrate different treatments. Note V'eed front end with inset head lamps.

Rough sketch by Reddig shows mid-size hardtop car with transparent roof.

Drawn in the 1950's by Bill Reddig, this sketch shows direction cars would take in the 1960's-lower, flatter body, quad head lamps.

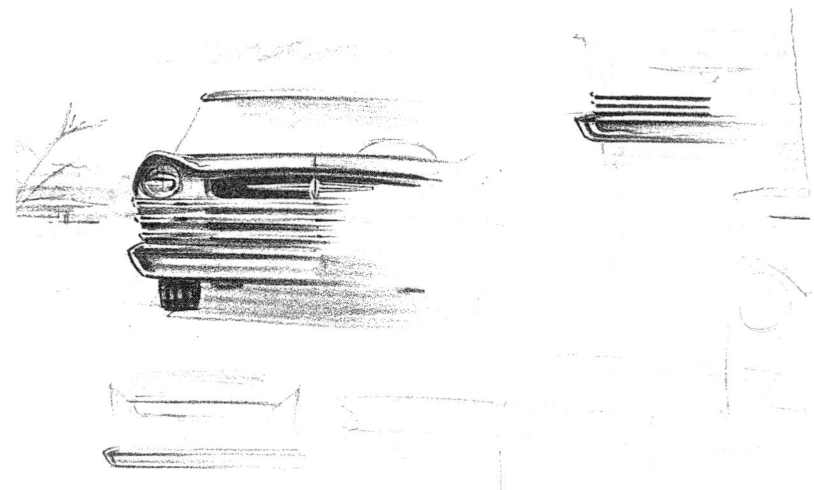

Another Reddig sketch, this shows a '50's style car with hooded head lamps, tail fins.

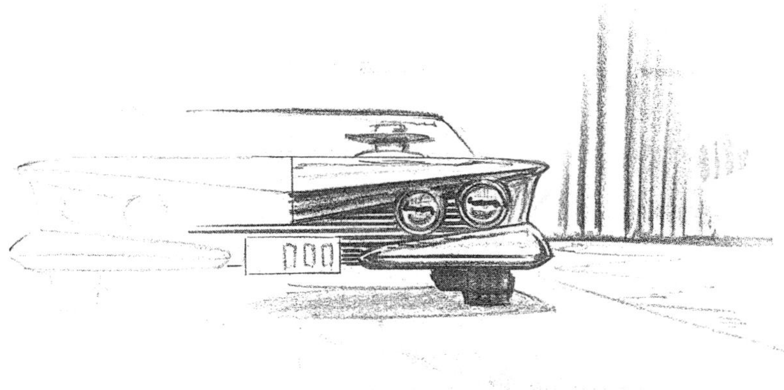

One problem designers had at the end of the 1950's was how to integrate new four head lamp systems into overall styling of cars. Here, Reddig comes up with a handsome modern idea.

Another try at a four head lamp system, this rather wild idea uses slanted pods to hold head lamps and turn signal lamp, very interesting grill shape.

Jack Garnier sketched out a modern looking Nash with open wheel wells, smooth fender contours. Inset in the front are small sketches of futuristic Nash cars. Interesting picture.

This idea, from Bill Reddig, shows a big Nash car styled for the future- head lamps under glass covers, fully skirted wheels, integrated bumpers, lots of window area.

If American Motors had continued the Nash brand into 1958 this is most likely what the Ambassador would have looked like. The product would have used a lengthened Rambler body with unique grille and trim, powered by the AMC 327 cid V8 engine. A very similar car dubbed the Ambassador by Rambler debuted instead.

About the Author

Just about every Nash, Hudson, Rambler, Metropolitan and AMC hobbyist is familiar with author Patrick R Foster's work. His articles have been published in many enthusiast magazines such as Special Interest Autos, Car Collector/Car Classics, AutoWeek, Collectible Automobile and Automobile Quarterly. Pat was the 1996 winner of the Society of Automotive Historian's prestigious **Carl Benz Award** for best periodical article in the field of automotive history. Two of his books "American Motors, The Last Independent" and "The Metropolitan Story" have won acclaim for their unique telling of automotive history. His upcoming new book "The Story of Jeep" is being eagerly anticipated by Jeep and AMC enthusiasts everywhere.

Pat resides with wife Diane, daughter Caitlin and Samantha the cat in the town of Milford, Connecticut, on the shores of Long Island Sound and can often be seen driving around town in his 1962 Rambler American convertible or his 1963 Rambler Classic.

Author Patrick R Foster with daughter Caitlin (left) and friend Elizabeth Ann. Photo courtesy W. Scott Cameron

Back Cover Illustration: Unique drawing by Nash designer Bill Reddig illustrates the different disciplines that go into designing a new car.